Survey and Analysis of Air Transportation Safety Among Air Carrier Operators and Pilots in Alaska

G.A. Conway, N.A. Mode, J.C. Manwaring, M. Berman, A. Hill, S. Martin, D.M. Bensyl, and K.A. Moran

U.S. Department of Health and Human Services

Centers for Disease Control and Prevention

National Institute for Occupational Safety and Health

November 2006

Ordering Information

Copies of National Institute for Occupational Safety and Health (NIOSH) documents and information about occupational safety and health are available from

NIOSH-Publications Dissemination
4676 Columbia Parkway
Cincinnati, OH 45226-1998

Telephone: 1-800-35-NIOSH
Fax: 513-533-8573
E-mail: pubstaft@cdc.gov
Web site: www.cdc.gov/niosh

This document is in the public domain and may be freely copied or reprinted.

Disclaimer:

Mention of the name of any company or product does not constitute endorsement by the National Institute of Occupational Safety and Health (NIOSH). In addition, citations to Web sites external to NIOSH do not constitute NIOSH endorsement of the sponsoring organizations or their programs or products. Furthermore, NIOSH is not responsible for the content of these Web sites.

DHHS (NIOSH) Publication No. 2007-102

Foreword

Aviation accidents are one of the leading causes of occupational fatalities in Alaska. Pilots in Alaska die at a rate nearly 100 times the mortality rate for all U.S. workers, and over five times the rate for all United States pilots. Unlike the rest of the country, many of Alaska's villages are not connected by a road system; commuter and air taxi operators serve as the main link between these villages and regional hubs, transporting people, cargo, and mail.

Although several federal programs have begun to address the issues surrounding aviation safety in Alaska, work remains to be done. This document describes a comprehensive survey of air-taxi operators and pilots in Alaska in which company and pilot demographics, flight practices, and attitudes about safety were examined. It provides information about current practices and how industry views potential safety measures, which is critical to designing effective prevention strategies.

The National Institute for Occupational Safety and Health, as the national agency responsible for occupational safety and health research, is committed to continuing to reduce the number of fatal occupational aviation crashes in Alaska. We look forward to further work with government, industry, and nonprofit partners who share our interest in protecting American workers who fly in Alaska.

John Howard, M.D.
Director
National Institute for Occupational Safety and Health
Centers for Disease Control and Prevention

Abstract

Because aviation crashes are one of the leading causes of occupational fatalities in Alaska, investigators at the Alaska Field Station of the National Institute for Occupational Safety and Health contracted with the Institute of Social and Economic Research at the University of Alaska Anchorage to administer two statewide aviation safety surveys, one of air carrier operators and one of active commercial pilots. Both surveys addressed pilot and company demographics; number of pilot flight hours (total, aircraft type, and instrument hours); flying experience in Alaska; and attitudes about safety, flying practices, and other salient risk factors.

Surveys from 153 commuter, air taxi, and public-use operators were received at a 79% response rate. Survey results were used to create an industry profile, compare operators' responses to their pilots' responses, and analyze and compare responses of operators with high fatal accident rates (designated "cases") to operators without high fatal accident rates (designated "controls").

Results indicated that the average case pilot had less career flight experience than the average control pilot and worked 10 hours a week more. Case operators were less likely to consider pilot fatigue a problem when scheduling flights and more likely to depend financially on timely delivery of bypass mail. Case pilots were three times as likely as controls to fly daily into unknown weather conditions. Nearly 90% of the case pilots reported that they never flew when so fatigued that they wanted to decline the flight, compared to 64% of control pilots. The findings suggest that the combination of pilot inexperience and longer work hours and work weeks may contribute to Alaska's high pilot fatality rate. Results of the operator-pilot comparisons suggest that financial pressures on operators may influence their views on what measures would be effective in preventing crashes. Many of the responses received in these surveys were consistent with the goals of three major, recently-implemented aviation safety programs in Alaska: the Medallion Foundation, the Federal Aviation Administration's Circle of Safety, and Capstone.

Contents

Foreword iii
Abstract iv
Abbreviations vi
Acknowledgments vii
1 Introduction 1
2 Methods 4
    2.1 Questionnaire Development 4
    2.2 Operator Sample 5
    2.3 Pilot Sample 6
    2.4 Survey Representativeness 7
    2.5 Data Analysis 8
        2.5.1 Estimates of Pilots and Flight Activity 8
        2.5.2 Characteristics of Operators and their Pilots 8
        2.5.3 Comparative Analyses 9
3 Results 11
    3.1 Workforce and Flight Activity 11
    3.2 Characteristics of Operators and Their Pilots 12
    3.3 Accident Prevention Measures 14
    3.4 Comparison of Large Operator and Pilot Responses 15
    3.5 Comparison of Operators With & Without High Fatal Crash Rates .. 17
4 Discussion 21
    4.1 Operators and Pilots – Activities, Practices and Perceptions 21
    4.2 Comparisons of Large Operators and Their Pilots 22
    4.3 Case and Control Comparisons 23
    4.4 Strengths, Potential Limitations, and Biases 24
5 Recommendations 26
6 Conclusions 27
References 29
Appendices 31
  A. Operator/Small Operator Questionnaire Summary 31
  B. Pilot Questionnaire Summary 43
  C. Additional Resources 55

Abbreviations

AKDOL	Alaska Department of Labor
CFR	Code of Federal Regulations
IMC	Instrument Meteorological Conditions
ISER	Institute of Social and Economic Research
FAA	Federal Aviation Administration
FAR	Federal Aviation Regulations (CFR Title 14)
FSDO	Flight Standards District Office
GPS	Global Positioning System
NIOSH	National Institute for Occupational Safety and Health
NTSB	National Transportation Safety Board
U.S.	United States
VIS	Vital Information System

Acknowledgments

The authors wish to thank Priscilla Wopat, Diana Hudson, Linda Bradford, Lance Kissler and Joshua Tabor for very helpful editing and formatting. The following University of Alaska Anchorage, Institute of Social and Economic Research staff assisted in the design, collection, and analysis of the survey data: Rosyland Frazier, Virgene Hanna, Patricia Deroche, and Darla Siver. This survey would not have been possible without their diligent efforts. The authors extend sincere thanks to Karen Casanovas and colleagues at the Alaska Air Carriers' Association, Felix McGuire and colleagues of the Alaska Airmen's Association, and Daniel Perry and George Kobelnyk of the Federal Aviation Administration, Alaska Flight Standards Division, for their helpful review of and suggestions for the survey instrument. The authors also wish to thank Nancy Stout and Timothy Pizatella of the NIOSH Division of Safety Research for their support throughout the project. Finally, the authors wish to express their gratitude to the pilots and managing personnel of the Alaska aviation operators who participated in these surveys.

1 Introduction

In 1999, the U.S. Congress funded the implementation of a federal initiative—the Alaska Interagency Aviation Safety Initiative—to reduce aviation-related injuries and fatalities and to promote aviation safety in cooperation with the air transportation industry and pilots in Alaska. This initiative is led by the National Institute for Occupational Safety and Health (NIOSH) in partnership with three other federal agencies that share an interest in preventing aircraft crashes and promoting aviation safety. These are the Federal Aviation Administration (FAA), the National Transportation Safety Board (NTSB), and the National Oceanic and Atmospheric Administration's National Weather Service.

This initiative complemented another, congressionally funded initiative to reduce aviation fatalities-the Capstone program, sponsored by the FAA. This joint FAA/industry effort includes installation of improved avionics in aircraft used in small commercial operations; improvements in ground infrastructure for weather information, data link communications, and flight information services; and development of new global positioning systems- (GPS-) based nonprecision instrument approaches at remote airports.

A disproportionate number of all U.S. aircraft "accidents"[a] occur in Alaska. Between 1990 and 2002, there were 434 commuter and air taxi accidents (CFR Part 135)[b] in Alaska—36% of all such accidents in the United States. The state with the next highest number of commuter and air taxi accidents was Florida, which had only 4% of all such U.S. accidents. Of the Alaska commuter accidents, 67 were fatal, resulting in 194 deaths (21% of all U.S. commuter and air taxi aviation deaths).[1]

Aviation accidents are a leading cause of occupational fatalities in Alaska. Between 1990 and 2002, aviation accidents in Alaska caused 130 occupational pilot deaths. This is equivalent to a rate of 385 deaths per 100,000 pilots per year, nearly 100 times the mortality rate for all U.S. workers (4 per 100,000 workers per year[2]) and over five times

---

[a] This document uses the National Transportation Safety Board definition of "accident" for consistency with other publications. An accident is an "occurrence associated with the operation of an aircraft which takes place between the time any person boards the aircraft with the intention of flight and until such time as all such persons have disembarked, and in which any person suffers death or serious injury, or in which the aircraft receives substantial damage."

[b] In this document, Code of Federal Regulations (CFR) refers to Title 14, Chapter I, which includes the Federal Aviation Administration and commercial aircraft in the United States. CFR Part 135 is also cited as "14CFR135" or Federal Aviation Regulations (FAR) Part 135.

the rate for all U.S. pilots (70 per 100,000 workers per year). Thus, an Alaskan pilot would have an 11% risk of death from an aircraft accident over a 30-year career.

These statistics may reflect some of the unique features of aviation in Alaska. Although more than half the population lives in the state's three major cities, many people live in remote villages. Commuter and air taxi operators serve as the main link between these villages and regional hubs, transporting people, cargo, and mail. In 1994, commuter airlines in Alaska served 238 locations, only five of which had road connections to the airline hub.[3] Approximately 85% of the aircraft are single-engine. These operations are a vital component of the transportation system in Alaska.[3]

Additional unique Alaska features affecting aviation are large areas of high, mountainous terrain; flat, marshy tundra; and an extensive coastline. These factors at Alaska's northern latitudes result in diverse climatic zones and associated weather that is variable and often harsh. Poor visibility and rapidly changing weather are common and contribute to the problems of air transportation in Alaska. Because of the high cost to cover such a large area, most of Alaska has been without usable infrastructure for instrument flight rule routes that permit low-altitude flying into small villages.

Due to Alaska's high accident rate, the FAA, NTSB, and other agencies have investigated many aspects of the regional airline industry. Most of these studies focus on accident report data.[4-10] A few have initiated surveys of pilots[3, 11-14] or audits of operators.[15] Studies based on accident reports describe common accident profiles. Several common fatality scenarios were identified, including take-off and landing errors and flying under visual flight rules into instrument meteorological conditions resulting in a controlled flight into terrain. Using data from pilot surveys, some papers compared pilots working for operators with high numbers of controlled flight into terrain or take-off/landing crashes to pilots working for other operators.[12, 13, 15] These studies led to recommendations for improved training, changing the safety culture, and providing better and more accessible weather information. Among the many changes that have occurred in Alaska was the installation of 16 remote video weather cameras during 1995-2000.

With one exception (NTSB 1995[3]), previous studies and the resulting recommendations have not addressed economic incentives that might put pressure on operators and pilots to fly in unsafe conditions. These incentives include overtime pay for pilots, income from bypass mail delivery, and pressure from passengers. Bypass mail is a federal subsidy for heavy shipments that would otherwise have to go as air cargo (at more expensive rates) to people and businesses in towns off the road system. Carriers contract with the U.S. Postal Service to deliver this mail and are required to deliver it by the end of the next business day after they receive it. If one carrier declines to deliver the mail, the task can be re-assigned to another carrier. While the U.S. Postal Service does allow for delays caused by bad weather, if another carrier chooses to fly into a given area, the mail goes to that carrier. So those who depend more on income from bypass mail may have more incentive to fly in marginal weather. The previous studies also did not consider the possible role of positive incentives such as insurance rate reductions for safe operators.

From August 2001 through January 2002, NIOSH sponsored two statewide air transportation safety surveys, one of Alaska commuter and air taxi operators (Appendix I: Operator/Small Operator Questionnaire Summary) and one of pilots (Appendix II: Pilot Questionnaire Summary) and contracted with the University of Alaska Anchorage, Institute of Social and Economic Research (ISER) to administer them. These surveys built upon previous investigations, providing an updated, in-depth survey of both pilots and their employers.

The intent of the surveys, as outlined in the original documentation, was to accomplish four tasks:
1. improve estimates of the professional Alaskan pilot workforce and flight activity;
2. identify the perceptions, activities, and practices of air carrier operators and pilots;
3. recommend prevention strategies based on survey results;
4. provide a basis to assess and evaluate the effectiveness of future interventions.

To this end, the surveys addressed current safety practices and training, as well as attitudes toward regulations and potential safety measures. Information about pilot and operator demographics, number of pilot flight hours (total, aircraft type, and instrument hours), number of hours of flying experience in Alaska, management and pilot attitudes about safety, flying practices, and other salient risk factors was also collected. Results were expected to provide valuable information to those developing interventions for reducing the incidence of commuter and air taxi crashes.

2 Methods

2.1 Questionnaire Development

To develop the questionnaires, NIOSH conducted focus groups from May through November of 2000 in five Alaska regions (Anchorage, Juneau, Bethel, Barrow, and Kotzebue). Pilots, operators, and community members participated in the focus groups. Themes emerging as concerns or barriers to aviation safety in Alaska were—

(1) Inadequate weather reporting and a lack of weather-reporting equipment and trained weather observers.
(2) Limited airport, airway, and navigation infrastructure.
(3) The very limited pool of experienced pilots in air taxi and small commuter operations resulting from the continuously high turnover of pilots. Once pilots become experienced in small operations, they typically seek and obtain jobs as pilots with major airlines and cargo carriers.
(4) A need for enhanced training in local conditions and more vigilant supervision of less-experienced pilots.
(5) A desire for increased support from federal regulators on safety-specific issues, including training and standardized interpretation and enforcement of regulations.
(6) Pressures—economic, passenger, and self-induced—to continue or take off in adverse and deteriorating weather conditions.

The first phase was to draft two preliminary questionnaires. In designing the pilot questionnaire, research staff took into account respondent sensitivity to questions about practices that are contrary to federal aviation regulations (FAR's). In addition to an understandable reluctance to admit to breaking the law, some pilots also raised concerns that their responses to such questions would be used for enforcement purposes. For these reasons, pilots were not asked questions about their employers that might call for explanations of practices or procedures contrary to aviation regulations.

The next phase was to pretest the questionnaires. Representatives from six companies pretested the operator questionnaire, and six pilots pretested the pilot questionnaire. Results from the pretest indicated which questions and terms were confusing, confirmed which questions were actually measuring perceptions and attitudes, and determined the amount of time required to administer the survey. Using this information, the final questionnaires were constructed.

2.2 Operator Sample

The operator survey population consisted of private companies, government agencies, and nonprofit organizations. Private companies which self-identified to the FAA as air transportation businesses carrying passengers and/or freight in Alaska as of November 2000 with aircraft having fewer than 10 seats were included. Government agencies and not-for-profit organizations included in the population operated public-use aircraft in Alaska as of November 2000.

ISER drew the sample from the FAA's Vital Information System (VIS) database, which the FAA uses to track its regulatory and licensing actions. ISER selected all companies supervised by the FAA Alaska region that were certified under CFR Part 135 (commuter airlines and air taxis) and grouped operators both by size and geography. The sample included commercial operators; federal, state and local public agencies; and one nonprofit, noncommercial corporation (the Civil Air Patrol). The only CFR Part 91 (general aviation) operations included in the study were those of government and CFR Part 135 operators flying under CFR Part 91. Lodge owners, guides, and similar professions were included only if they had a CFR Part 135 certificate.

ISER grouped operators into "large" (operators employing three or more pilots) and "small" (operators with one or two pilots). Large operators were then grouped into a case group and a control group (see below) based on the estimated number of annual fatal accidents per pilot during the portion of the period during January 1990 to June 2001 when the operator maintained CFR Part 135 operations in Alaska.

According to the VIS, 123 Alaska operators employed three or more pilots; 285 operators employed one or two pilots. Most commercial operators in Alaska employ just one or two pilots; however, larger operators account for the most flight hours in Alaska. To get information from operators and pilots having the most flight hours, as well as address the diversity across Alaska, ISER attempted to survey all the large operators and one-third of the small operators. Geographically, the FAA's Anchorage Flight Standards District Office (FSDO) supervises 78% of all small operators (covering south-central and southwest Alaska) along with the Fairbanks and Juneau FSDO (covering interior and northern Alaska and southeast Alaska, respectively) supervising 11% each. To ensure geographic representation that captured the variations across the state in weather, terrain, remoteness of destinations, and aviation infrastructure, the small-operator sample was stratified by the supervising FSDO. A random sample of 60 small Anchorage operators (about 28%), 18 Fairbanks operators (56%), and 16 Juneau operators (53%) was selected (Table 1). The Fairbanks and Juneau regions were over-sampled relative to the Anchorage region due to the small total number of operators in those regions. The survey response rate for all operators was 79%. Company identities were kept confidential and unknown to the government employees involved in conducting the study.

Methods

Table 1. Number of operators and survey response rate for large and small Alaskan commuter, air taxi and public aircraft companies

	Large Operators	Small Operators			Total
		Anchorage	Fairbanks	Juneau	
Original operator sample	123	60	18	16	217
Revised operator sample (still in business in Alaska)	116	49	17	12	194
Nonresponses	20	9	6	6	41
Completed operator surveys	96	40	11	6	153
Operator response rate	83%	82%	65%	50%	79%

2.3 Pilot Sample

The pilot survey targeted pilots currently employed by operators responding to the operator survey. The pilot sample was generated from interviews with the air carrier operators. Pilots were randomly selected from the large companies which participated and agreed to let their pilots be surveyed. ISER asked the company to provide sampling data and contact information for their pilots in one of the following ways:

Most preferred: The company provides names and contact information for a complete list of their pilots so that ISER could contact the pilots directly;

Next most preferred: The company provides a list of pilot names from which to draw a sample and would distribute survey forms to the selected pilots; and

Least preferred: The company follows instructions on how to draw a sample of their pilots and then distributes survey forms to the selected pilots.

For small companies, the pilot sample was identical to the operator sample, that is, operators were also pilots. For small operators, the operator and pilot questionnaires were combined into one document and duplicate questions were eliminated. Combining the questionnaires reduced costs and the time burden on the small-operator respondent by obtaining both operator and pilot information in one contact. Because of discrete sampling of pilots, the actual sampling fraction varied by the size of the operator. The maximum sampling fraction was 40% (two pilots in a five-pilot company), the minimum sampling fraction was 22%, and the average was 25% overall. Of the 88 large operators which allowed pilots to receive surveys, 75 companies (85%) had at least one pilot complete a survey. A total of 204 individual pilot surveys were completed from the possible sample of 295 pilots working for large operators, a response rate of 69%.

A consent form was sent with the questionnaires advising respondents of the confidentiality of the information they were providing. It also included information on the authority and purpose for data collection and told respondents that their participation was voluntary, that responses would not be used in enforcement actions against them, and that the survey results would be made available to the air carrier operator and pilot associations, federal agencies, and other interested parties in a summary format only, without any personal or corporate identifiers.

ISER interviewed operators from August 2001 through January 2002 and pilots from December 2001 through February 2002. Surveys were mailed to all selected companies and followed up by telephone and fax as necessary. In cases when telephone contact was unsuccessful or when the operators preferred, interviewers completed the interview in person.

2.4 Survey Representativeness

Because the survey was stratified, the results needed to be weighted to properly represent the characteristics and attitudes of Alaska CFR Part 135 operators as a whole. Since the population numbers from the VIS database included operators no longer in business in Alaska, ISER first calculated adjusted population numbers. VIS totals were adjusted based on the numbers of sampled companies that had changed strata or gone out of business. The resulting estimates of operators in each stratum (Anchorage, Fairbanks and Juneau) are not integers (Table 2). To calculate weights, ISER divided the total number of small operators in each stratum by the number of completed interviews in that stratum. Each completed interview in Anchorage represents over four operators, each interview in Fairbanks represents 2.59 operators, and each interview in Juneau represents 3.75 operators.

Table 2. Weighting small operator survey data to represent all small operators

	Anchorage	Fairbanks	Juneau
Total estimated small operators	178.4	28.4	22.5
Total completed surveys	40	11	6
Weight	4.46	2.59*	3.75

* Actual calculation 28.444/11 = 2.585

Pilot weights were calculated by dividing operator weights by the fraction of pilots interviewed. A separate pilot weight was calculated for each company. For a technical discussion of weighting in this study, refer to Conway et al., 2004.[16]

Possible response bias was assessed using variables for which information existed for nonrespondents as well as respondents. Available test variables included company size, location, and number of accidents. Lack of information on unscheduled flight hours precluded tests for a bias in the accident rate, especially for small operators for whom unscheduled activities make up a large fraction of operations. A proxy for accident rate was calculated by dividing the number of accidents between January 1990 and June 2001 by the number of pilots reported in the VIS. No significant relationship was found between this rough accident rate and whether operators were willing to complete the survey. Likewise, size and location were not associated with a greater likelihood of response or refusal/noncontact. Although Juneau's response was lower than in other areas, the total number of small operators in Juneau was too small for the lower rate to be statistically significant.

2.5 Data Analysis

In accordance with the intent of the survey instrument, the survey data were used to: create estimates of numbers of pilots and flight activity, identify characteristics of operators and their pilots, and perform comparative analyses to identify possible prevention strategies.

2.5.1 Estimates of pilots and flight activity

Estimates of the number of pilots working for commuter, air taxi, and public agency operators in Alaska were derived from questions in the operator survey about the total number of pilots employed, and the total flight hours. Questions about pilot numbers were specific to season, while the flight hours included all those flown during 2000. Estimates and the corresponding 95% confidence intervals were calculated using the stratified sampling weights (Table 2) based upon sampling theory (for example see Thompson 1992[17]).

To estimate the total number of pilots employed in Alaska, Alaska Department of Labor (AKDOL) data were used to obtain the number of pilots working for companies that had not been included in the sampling frame. AKDOL data, drawn from unemployment insurance records, include all pilots who work for companies that employ 20 or more people (total, not just pilots) in Alaska. What CFR Part governs a pilot is not relevant to their inclusion in AKDOL data. It was, however, key to selecting the appropriate sample for the survey.

The survey focused on air operators flying under CFR Part 135 and public agencies. As a result, some pilots were counted in one data source, some in the other, some in both, and a few in neither. For example, this survey did not include pilots flying exclusively for large airlines (Alaska Airlines, Northwest Airlines; CFR Part 121). ISER, as the contractor, was able to view the company names, match the survey data with the AKDOL data, and identify pilots not included in the survey. The estimated total number of CFR Part 135 pilots employed during the peak summer season was combined with the pilots described in AKDOL who had been excluded from the original survey.

2.5.2 Characteristics of operators and their pilots

Survey results about operator and pilot characteristics were summarized either by reporting the percentage responding positively (for example, percentage holding an instrument license), or by the average value (mean) of all responses (for example, the average number of years of each pilot's flight career). The median, or value such that half the values are larger and half the values are smaller, was also presented for results on pilot flight experience. If the mean and median are similar in value, then none of the responses were substantially higher or lower than the average. If they differ, then some of the responses were very different than the rest and the median is a better representation of the majority of responses.

The stratified sample meant that ISER staff spoke to almost all the larger operators in the state, but only about one in five of the smallest operators. To reflect the aviation community more accurately, the small operator results were weighted to account for the same two-thirds of all operators in the weighted sample that they represent in the total population. Not all operators or pilots surveyed answered all the questions and results represent the average or percentage of all the answers received.

2.5.3 Comparative analyses

Survey results were compared in two ways, one in which responses of large operators were compared to the average response of their pilots to ascertain similarities and differences (particularly with respect to attitudes and beliefs), and one in which responses of operators and pilots of companies having high rates of fatal crashes (designated "cases") were compared to responses of operators and pilots of companies not having high fatal crash rates (designated "controls"). Both analyses were restricted to survey responses from only the large operators.

In the first analysis, survey responses of large operators and their pilots were summarized. Questions about attitudes, beliefs, and preferences had been designed with Likert scale responses. For example, the possible answers to a question about risk reductions measures were: "very effective", "somewhat effective", and "not effective". While one could in principle compare qualitative responses directly, we chose instead to compare means of ordinally scaled variables. The principal reason for comparing means is to obtain a clear indicator of the direction of response differences. For example, group responses for a question on perceived effectiveness might show that 50% of the respondents perceived that a measure was not effective at all (0 on the Likert scale) and 50% perceived the measure as very effective (3 on the scale). These values would be a significant difference in a chi-square test from a group where all respondents perceived that the measure was somewhat effective (2 on the scale), but it would not show a clear direction for the difference.

Statistical tests were conducted at the operator level to examine differences between operators' responses and the average response of their pilots in paired sample tests. Paired tests take into account the potential similarity of views between operators and the pilots they have hired. A significance level of 0.05 was established for the difference of means tests (t-tests). As a result of the large number of tests performed, the overall or experiment-wise error rate will be higher than 0.05, and a few tests could be significant due to chance. As with most surveys, it is the pattern of results that may be most meaningful. Since many of the questions on the same subject had to be worded differently for pilots and operators in order to make sense in different contexts, these statistical tests address differences in questions that are related but not identical. Only the responses of operators and pilots from large operations were analyzed since the pilots of small operations were speaking for the company as well as themselves, often because they were the sole pilot, so logically there was no difference between small operators and the pilots working for small operators.

In the second analysis, data on large operators were used to understand if practices and attitudes differed between operators with a higher number of fatal crashes than expected (cases) and the rest of the operators (controls). Fatal accident data were gathered using the publicly available NTSB Aviation Accident Database. Companies that had been in business longer and had more exposure were expected to have had more fatal accidents. An operator was classified as a case if their crash rate was higher than the expected number of fatal crashes given the number of pilots employed. A probability of 0.30 was used to divide cases and controls. An operator fell into the case group if there was less than a 30% probability of observing the operator's number of fatal crashes compared to the expected number. The expected number of fatal crashes assumes that the underlying crash rate was the same as the average Alaska operator. Small operators were excluded from this analysis because they generally had insufficient flight activity to determine reliable crash rates, and pilots were assigned based on their employers' crash rates. For a detailed discussion of case and control assignment, refer to Conway et al., 2005.[18] Differences between cases and controls among pilots in terms of their responses to survey questions were analyzed using Pearson chi-square($\chi 2$) and difference of means tests (t-tests) with a 0.05 significance level.

3 Results

3.1 Workforce and Flight Activity

Information for sampling was based on the number of pilots reported as employed in the November 2000 VIS database; for analysis, the operators were classified based on their responses to the question, "How many pilots do you currently employ?" A few companies fell into one category based on the VIS, but the other in their survey answer (Table 3). Survey answers were assumed to be more up-to-date than the information available in the VIS, and so all analyses were based upon the number of pilots reported on the survey. Fourteen companies sampled as large were analyzed as small; three companies sampled as small were analyzed as large.

Table 3. Comparison of large and small operator classifications using the FAA Vital Information System to create the sample, and survey data to establish groups for analysis

		Classification according to VIS (sample)		
		Large	Small	Total
Classification according to survey (analysis)	Large	82	3	85
	Small	14	54	68
	Total	96	57	153

The survey asked how many pilots each company typically employed in the fall, winter, spring, and summer seasons. Based on responses, an estimate of statewide employment at the time of the survey as well as seasonal employment was generated. The employment numbers at the time of the survey largely reflected fall employment, since the interview dates ranged from August 2001 to January 2002. Employment was highest in summer and lowest in winter (Table 4). Estimates of employment by CFR Part 135 operators and public agencies ranged from 1,426 pilots in winter to 1,907 in summer. The number of pilots employed by individual companies ranged from 1 to 105. Over half of all companies employed only one pilot, and two-thirds employed only one or two pilots; only 10% employed 10 or more pilots. About 95% of the pilots were employed by the one-third of the companies that had more than two pilots. The average number of pilots employed by companies was 1.7 pilots in 1999, 1.9 in 2000, and 1.8 in 2001. Survey data indicate that in 2000, Alaska CFR Part 135 and public agency operators statewide flew an estimated total of 420,000 scheduled flight hours (95% CI: 275,000-565,000) and 415,000 unscheduled flight hours (95% CI: 370,000-460,000).

Results

Table 4. Estimated number and 95% confidence interval of pilots employed by Alaska-based CFR Part 135 operators and public agencies

	Lower 95% CI limit	Estimate	Upper 95% CI limit
Employed at time of survey	1,671	1,856	2,041
Summer employment	1,731	1,907	2,116
Winter employment	1,247	1,426	1,631

The final estimate of all pilots employed in Alaska during the peak season, incorporating the numbers of pilots flying for other types of aviation operators from the AKDOL, was 2,742 (95% CI: 2,551-2,932). This estimate excluded people who flew for companies that were both very small and not regulated under CFR Part 135. Typical examples are fish spotters, pilots providing "incidental transportation" for lodges under CFR Part 91, and some flight instructors.

3.2 Characteristics of Operators and Their Pilots

A higher percentage of large operators surveyed had formal programs to implement risk reduction measures, including training and supervision measures than small operators surveyed (Table 5). Eight of the large operators also had CFR Part 121 certificates, while none of the small operators did. Nearly all companies permitted pilots to cancel flights. However, only 30% of the large operators and 19% of the small operators required higher than the minimum regulatory weather conditions for flying, and only 12% of the large operators and 4% of the small operators had a written list of launch conditions.

Table 5. Characteristics of large and small Alaskan commuter, air taxi, and public agency operators

	Large operators	Small operators
Number of respondents	85	68
Operator characteristics		
Mean flight hours 2000	5,507	766
Median flight hours 2000	2,739	500
Mean percentage increase in insurance costs in past 18 months	39	15
Median percentage increase in insurance costs	20	15
Percentage of operators who pay pilots overtime	33	0
Risk reduction measures in place (%)		
Higher-than-FAA weather minimums required	30	19
Whiteout pilot training	50	16
Low-visibility pilot training	60	19
Flat-light pilot training	52	15
Recovery from IMC pilot training	66	19
Whiteout pilot check rides	42	11
Low-visibility pilot check rides	49	14

IMC = Instrument meteorological conditions.

Table 5 (*continued*). Characteristics of large and small Alaskan commuter, air taxi, and public agency operators

	Large operators	Small operators
Flat-light pilot check rides	38	11
Recovery from IMC pilot check rides	53	15
Written list of launch conditions	12	4
Pilots can cancel flights	100	99
Other employee can cancel flights	88	26
Outside person can cancel flights	18	6

IMC = Instrument meteorological conditions.

Essentially all pilots surveyed had commercial and instrument ratings (93-100%). These ratings are required for pilots flying for CFR Part 135 operations, but not for those flying for public agencies. Many pilots are more highly qualified than required; 72% of the large operator pilots and 42% of the small operator pilots had either Certified Flight Instructor or Airline Transport Pilot certificates (Table 6).

More pilots working for large operators had multiengine land ratings than did small operators and the reverse was true for single-engine sea ratings (Table 6). Some of the measures of pilot experience–flight hours, years of experience, year-round Alaska experience–had similar means and medians, especially among the small operators' pilots. In those cases, the mean or median fairly represented the years of experience of many pilots, with a few having substantially more or less experience than average. However, median hours of instrument flight experience (Alaska and total) were well below the means, indicating that many pilots had few or no such hours, and a few had many hours. Although almost all pilots were instrument-rated, 16% had zero hours of instrument flight in Alaska.

Table 6. Characteristics of pilots by size of Alaskan commuter, air taxi, and public agency operators

	Large operators	Small operators
Number of respondents	197	64
Pilots holding various pilot certificates/privileges (%)		
Commercial	96	100
Instrument	95	93
Airline transport pilot	64	35
Helicopter	18	9
Flight instructor	34	17
Pilots holding various aircraft ratings (%)		
Single-engine land	86	81
Multiengine land	70	43
Single-engine sea	56	73

Results

Table 6 (*continued*). Characteristics of pilots by size of Alaskan commuter, air taxi, and public agency operator

	Large operators		Small operators	
Multiengine sea	15		20	
Other	18		6	
Female pilots (%)	6		2	
Mean age (years)	42		49	
Mean number of employers (as pilot)	4.3		3.6	
Average work day (duty hours/day, peak season)	12		11	
Average work week (duty hours/week, peak season)	72		67	
	Mean	Median	Mean	Median
Length of flight career (years)	15	13	20	20
Year-round Alaska flight experience (years)	10	6	5	6
Flight hours, total	9.408	7,034	10,918	10,006
Flight hours in last 12 months	636	600	482	400
Flight hours, Alaska	7,084	5,044	9,731	9,427
Flight hours, Alaska in last 12 months	595	600	457	400
Instrument hours, total	1,096	337	690	110
Instrument hours, Alaska	797	150	284	50

Pilots had been flying for an average of 16 years, from less than 1 year to 50 years. Two-thirds of the pilots working for large operators had flown year-round, rather than seasonally, for 1 to 40 years (mean of 9.6 years) throughout their Alaska careers. Of those with seasonal flight experience, the mean was 4.5 years of Alaska experience. Over their entire careers, pilots had worked for a mean of 4.2 companies. Pilots are typically male (95%); 60% are in their 30s and 40s. Most (84%) have education beyond a high school diploma; 45% have a bachelors or higher degree.

Most pilots appear to work long duty hours during the busy summer season (numbers include all work time, not just flight hours). The average reported work day and work week were 12 and 71 hours, respectively. Over 86% of the pilots reported that they worked more than 50 hours a week during the busy season.

Pilots for both large and small operators perceived their jobs as being as safe or safer than other jobs. When asked whether a pilot's job was more dangerous than other jobs, among pilots working for large operations, 9% said much safer, 8% slightly safer, 31% as safe, 44% slightly more dangerous, and 7% much more dangerous. Pilots for small operators were even more optimistic: 8% much safer, 13% slightly safer, 52% as safe, 21% slightly more dangerous, and 6% much more dangerous.

3.3 Accident Prevention Measures

Operators and pilots generally agreed that improved weather information, especially via video camera, and weather reporting by and consultation with trained weather observers; improved decision-making policies and skills; and regional hazards training about local weather and terrain would all be effective ways to prevent crashes (Table 7). Not much optimism was shown about improving passenger understanding of weather (that is, passengers bringing pressure to fly into poor weather or visibility) as a successful prevention strategy. Neither was there much enthusiasm expressed by any of the groups for changes to the current allocation and management system for bypass mail (commercial pilots carrying U.S. mail to remote villages).

Table 7. Percentage of operators and pilots rating various accident prevention measures as "very effective"

	Large operator management	Small operators	Pilots for large operators
Weather information			
More locations with staffed weather reporting	79	82	76
Increased accuracy of existing weather reporting	77	77	76
More locations with automated weather reporting	64	59	75
Increased use of weather video cameras	62	75	75
Improved passenger understanding of weather hazards	33	34	18
Training			
Decision-making training	75	68	79
Regional hazards training	70	68	82
Whiteout/flat-light training	48	60	73
Meteorology training	36	48	55
Rewards and incentives			
Financial incentives for operators with no accidents or incidents	50	39	23
Salary-based pay	34	28	42
Pilot rewards for flights or flight hours without accidents or incidents	26	23	27
Pilot experience			
Better checks of a pilot's flying history before hiring	42	28	20
More flight time for new pilots	39	40	41
Other			
More time to deliver bypass mail	33	9	13
Written criteria for go/no-go decisions	30	31	37
Changes in how bypass mail is given to operators	25	10	11

3.4 Comparison of Large Operator and Pilot Responses

The responses of large operators were compared with the average responses of their pilots for an operator-level analysis (Table 8). Results show paired tests of the mean difference between operator and pilot responses within the same company.

Results

Table 8. Difference between large operator and pilot responses to similar questions

Survey Question	Operator-pilot mean*	t-statistic	p-value
Attitudes toward FAA oversight			
Regulations interfere with getting the job done	-0.05	-0.54	0.59
Higher-than-FAA weather minimums sometimes needed	-0.53	-7.79	**0.00†**
Company provides formal pilot training			
Whiteout	-0.20	-2.48	**0.02**
Low visibility	-0.18	-2.44	**0.02**
Flat lighting	-0.17	-2.22	**0.03**
Recovery from IMC	-0.13	-1.72	0.09
Perceived effectiveness for preventing crashes			
Improvements in meteorology pilot training	-0.20	-1.77	0.08
Improvements in decision-making pilot training	-0.02	-0.28	0.78
Improvements in whiteout pilot training	-0.20	-1.99	**0.05**
Improvements in regional hazards pilot training	-0.13	-1.70	0.09
Better checks on pilot's history	0.31	2.74	**0.01**
Better passenger understanding of weather hazards	0.27	2.38	**0.02**
Changes in the way bypass mail is given to operators	0.38	2.75	**0.01**
More time for bypass mail	0.51	3.68	**0.00†**
Financial incentives for operators with no accidents	0.54	4.35	**0.00†**

* Positive values indicate that on average operators agreed with the statements more than their pilots. Negative values indicate that on average pilots agreed with the statement more than operators.
† Indicates p-value less than 0.005.
Bold values indicate p-value ≤ 0.05.

Many pilots believed that some routes needed higher-than-FAA weather minimums to keep an adequate margin of safety. Pilots reported receiving training for flying in adverse weather conditions, but also stated that their companies have no written training program; that is, training programs are informal. On effectiveness questions, pilots generally viewed additional training to be more effective in preventing crashes than did operators. Companies viewed better preemployment hiring checks to be more effective than did pilots. These results appear to reflect the differing perspectives of the two parties. Pilots viewed better weather reporting as more effective, but differences between support for additional automated weather stations and additional video cameras were not significant. On the other hand, pilots were less enthusiastic than operators about better passenger understanding of weather hazards, changes in bypass mail policies, and financial incentives for safety. These are all measures that have a direct effect on company finances. Pilots were not as optimistic as operators that financial incentives for operators who did not have accidents would result in improvements in safety.

In addition to the differences analyzed, differences between operator and pilot perception of fatigue may exist. Among large operators, only 6% responded that fatigue was a major problem in scheduling, 46% responded that it was a minor problem, and 48% responded that it was not a problem. Pilots were asked how often during the peak season they would have preferred to decline a flight due to fatigue, but flew anyway. Among pilots for large operators, 15% had made such a decision weekly during the peak season, 7% monthly, 24% less than monthly, and 54% never made such a decision. Among small operators, the respective proportions were 13% weekly, zero monthly, 25% less than monthly, and 63% never. The questions about fatigue for operators and pilots are not strictly comparable, however, so a difference of means would not be a meaningful statistic.

3.5 Comparison of Operators With and Without High Fatal Crash Rates

Of the 85 large operators responding to the survey, 14 were categorized as high fatal crash operators (cases) versus another 67 large operators (controls) (Table 9). Four large operators who responded to the survey did not have business status information available for the entire period (January 1990 to June 2001) and thus were excluded from this analysis. The case operators generally flew more hours than controls and were less likely to consider pilot fatigue a problem when scheduling flights. Case operators were more likely to consider on-time bypass mail delivery important to their financial success. Overtime pay and changes in insurance rates were not significantly different between the two groups. The results show no significant differences between the risk reduction measures undertaken by the case and control firms; however, of the 13 listed policy risk reduction measures, only one was less common in cases.

Using information from the VIS database, ISER was able to examine the types of operators placed into the case and control groups. Eight operators in the control group held CFR Part 121 certificates as well as CFR Part 135 certificates as of June 2001, while none in the case group held CFR Part 121 certificates. All the public operators participating in the survey were in the control group. Operators in the case group were more likely to provide scheduled service according to their certificates (57% of cases versus 33% of controls), although this difference diminishes when all operators reporting scheduled flights in 2001 are included (57% cases versus 42% controls). Control operators reported having slightly more aircraft certified (under any CFR certificate) than case operators, although the difference was not statistically significant (a mean 14.4 aircraft per operator versus 13.1, $t = 0.4$, $p>0.05$). The fleet composition of case and control operators differed; case operators had many more of their fleet in single-engine land aircraft (72% of passenger aircraft reported by case operators versus 32% of control operator passenger aircraft) while control operators had more helicopters (30% of passenger aircraft reported by control operators versus 4% of case passenger aircraft).

Results

Table 9. Comparison of characteristics between operators having a high number of fatal crashes and other operators

	High fatal (cases)	Other (controls)	Significance (t or χ^2)
Sample size	14	67	
Operator characteristics			
Mean flight hours, 2000	8,468	5,577	**0.04**
Mean increase in insurance costs past 18 months (%)	40	35	0.80
Pay pilots overtime (%)	29	32	0.81
Pilot fatigue not a scheduling problem (%)	71	42	**0.05**
On-time mail delivery important to financial success	85	53	**0.04**
Risk reduction measures in place (%)			
Higher-than-FAA weather minimums required	29	29	0.99
Whiteout pilot training	71	45	0.07
Low-visibility pilot training	71	59	0.37
Flat-light pilot training	69	48	0.16
Recovery from IMC pilot training	71	62	0.49
Whiteout pilot check rides	50	42	0.59
Low-visibility pilot check rides	57	50	0.63
Flat-light pilot check rides	46	39	0.64
Recovery from IMC pilot check rides	57	50	0.63
Written list of launch conditions	14	12	0.81
Pilots have ability to cancel flights	100	100	1.00
Other employees can cancel flights	100	91	0.24
Outside person can cancel flights	14	23	0.48

Bold values indicate p-value ≤ 0.05.

Although few differences appeared between case and control operator survey responses, responses from pilots revealed significant differences in operations (Table 10). No pilot respondents employed by case operators and only seven pilot respondents of the control operators were female. The small number of female pilots overall resulted in low statistical power for testing gender effect differences between the groups. Pilots flying for case operators had one-third fewer years of flying experience, approximately half as many instrument hours overall, and half as many instrument hours in Alaska. On average, pilots flying for case operators had half as many hours with their current employer as control pilots. On the other hand, case pilots had worked significantly more hours in the past 12 months than other pilots, suggesting they were flying more hours in their current jobs.

Differences in responses to questions about working conditions confirmed the differences in working conditions between pilots of case and pilots of control operators. Case pilots worked more hours per day and more days per week on average than controls. Pilots of case operators worked nearly 13 hours per day and 81 hours per week during peak season, averaging 1 hour per day and 10 hours per week more than controls (includes all work time, not just flight hours). Paradoxically, nearly 90% of these pilots reported that they never flew when so fatigued that they wanted to decline the flight (corroborating the finding that their operators did not view pilot fatigue as a scheduling problem), compared to 64% of pilot controls. Although three times as many case pilots (18%) reported that they decided to fly into unknown weather every day, there was no significant difference between case and control pilots in their likelihood of perceiving their job as more dangerous than other jobs. All pilots of cases reported relying on headquarters or hub personnel, and station managers or other personnel at their destinations for the decision to launch a flight.

Table 10. Test results of selected characteristics and working conditions of pilots employed by operators having a high number of fatal crashes and other operators

	Pilots, high fatal (cases)	Pilots, other (controls)	Significance (t or χ^2)
Sample size	28	146	
Pilot characteristics			
Female	0	7	0.22*
Mean age, years	39	43	0.12
Mean flight career (years)	11	16	**0.02**
Mean flight hours, total	7,319	9,538	0.16
Mean flight hours in last 12 months	804	639	**0.02**
Mean flight hours in Alaska	5,881	6,713	0.55
Mean flight hours in Alaska in last 12 months	756	557	**0.00†**
Mean instrument hours overall	413	912	**0.01**
Mean instrument hours in Alaska	291	578	**0.04**
Mean number of employers (as pilot)	4	4	0.85
Total hours with current employer	1,806	3,732	**0.00†**
Working conditions			
Average hours per work day (busy season)	13	12	**0.00†**
Average hours per work week (busy season)	81	71	**0.00†**
View job as more dangerous than other jobs (%)	48	45	0.73
Never flew when fatigued and wanted to decline flight (%)	89	64	**0.01**
Decides to fly into unknown weather daily (%)	18	6	**0.04**
Use hub/HQ personnel before launching (%)	100	86	**0.04**
Use station manager/other personnel at destination(s) before launching (%)	100	88	**0.05**

* Fisher's exact test was employed due to small numbers.
† Indicates p-value less than 0.005.
Bold values indicate p-value ≤ 0.05.

One reason why pilots of case operators reported flying significantly more hours per day and per week than pilots of control operators might be that the risk of accidents is proportional to hours flown, and the greater flight activity of case pilots simply gives them more exposure to the risk of accidents. For this explanation to be correct, the assignment of operators to cases and controls would have to be biased toward placing operators with more pilot hours per day and per week into the case group, given each operator's underlying accident risk.

The evidence for such an alternative explanation of the results was examined by testing whether pilot hours per week predicted an operator's case-control status after controlling for total hours flown (exposure). To perform this test, a binary probit equation was estimated in which the dependent variable was case or control status based on fatal crashes per pilot, and independent variables were accidents per total hours flown and average survey pilot hours per week. If using accidents per hour flown to divide operators into cases and controls would have yielded similar results to using accidents per pilot (except for the difference in power), then the coefficient for accidents per hour flown in the estimated equation would be positive and significant. If using accidents per pilot biased the case-control selection toward firms with greater pilot hours per week, then hours flown per week would have a significant positive coefficient.

Results of estimating this probit equation showed no evidence of bias in the assignment of firms with higher pilot hours per week to the case group. The equation predicted case or control status correctly 97% of the time, with fatal accidents per hour flown positive and significant (Wald chi-square test $p<0.01$) and hours flown per week near zero and insignificant (Wald chi-square test $p=0.94$). This provides strong evidence that using the number of pilots instead of total hours did not bias the assignment of operators.

4 Discussion

4.1 Operators and Pilots – Activities, Practices and Perceptions

Operators and pilots both strongly supported improving meteorological services and consultation, including weather prediction, reporting, and deployment of more video weather cameras in critical locales, and increased numbers and involvement of trained weather observers. While there are on-going increases in the deployment of weather cameras and constant refinement of weather prediction and Web-based access to these predictions and current conditions, funding and administrative support for weather observers and Flight Service Station personnel has been controversial in Alaska for some time. Narrative responses to open-ended portions of the questionnaires and discussions in focus groups reflected a wide distrust of the accuracy of automated weather observation (beyond the direct form afforded by the weather cameras) and a strong desire to be able to talk with someone at a destination and in communities en route to consult on current conditions, near-term weather prediction, and advisable and/or best routes. The National Weather Service has been responsive to some of these concerns with the development of real-time weather consultations between pilots and personnel via the "mike-in-hand" program (http://www.alaska.faa.gov/at/notices/WX.htm).

The strong support expressed by both operators and pilots for improvements in and wider utilization of training in decision making (particularly for visual flight rules into marginal and instrument meteorological conditions), flat light, whiteout conditions, and regional hazards should facilitate the implementation of such measures. Support for more supervised flight time for pilots, particularly in their own region, and better checks of pilots' preemployment flying history by air carrier operators indicate that such changes should also be reasonably well received. There was also detectable support, though not as strong, for written criteria for go/no go decisions, pilot rewards for safe flying, and improved passenger understanding of weather hazards; thus, these interventions, if pursued, should be implemented more cautiously, with active consultation and collaboration with industry leaders and pilots.
Strategies involving changes in time allowed for delivery and how service is allocated for bypass mail were the least popular. While the majority of large operators expressed some enthusiasm for affording more time for delivery of bypass mail, and a slim majority supported changes in how it is allocated, neither of these measures were supported by smaller operators. Pilots working for larger operators only deemed increasing delivery time as likely to be "somewhat effective," while the majority

responded that allocation changes would not be effective. In general, the results of operator-pilot comparisons suggested that the differing financial pressures and incentives on operators and pilots may influence their views on what measures would be effective in preventing crashes. Results indicated that a respondent's position in a company (operator or pilot) was more influential in determining their responses than the company they worked for.

Large and small operators differed in their ability to provide procedural and operational risk-reduction measures. A higher percentage of large operators had formal programs to implement risk-reduction measures, including pilot training and check rides. Several of the large firms also operated under CFR Part 121; these firms also employed the vast majority of pilots in the state. The larger, more diverse operators would likely be more able to provide formal training procedures to their pilots. Responses from pilots from large and small firms were similar with regard to total flight hours, although pilots varied greatly in their instrument experience. Almost all pilots had the ability to cancel flights.

A consistent finding from both the operator and pilot surveys was the high intensity of work during the peak season. The average reported work day (duty hours, not flight hours) for a pilot was 11.5 hours, and the reported work week was 71 hours. Over 86% of the pilots reported that they worked more than 50 hours a week during the busy season. Fatigue can occur from disrupted sleep due to changing schedules, as well as a lack of sleep, and cumulative sleep loss can lead to impaired performance and diminished alertness.[19]

4.2 Comparisons of Large Operators and Their Pilots

In opinions expressed about the likely effectiveness of interventions, pilots viewed additional training as more effective than did operators, while operators often viewed preemployment hiring checks to be more effective than did pilots. These differences appear to reflect the differing perspectives of the two parties. Similarly, pilots viewed better weather reporting as more effective, but were less enthusiastic than operators about better passenger understanding of weather hazards; changes in bypass mail policies and financial incentives for safety also seemed to reflect differing perspectives.

Operators and pilots had different perceptions regarding fatigue. Only 6% of the large operators perceived that fatigue was a major problem in scheduling, while the pilots working for these firms indicated that fatigue during the peak season was more of a problem, and 22% responded that they made a decision to fly when fatigued either weekly or monthly. For small operators, where the operator might be the only pilot, a small percentage (13%) of these operator/pilots responded that they made a decision to fly when fatigued on a weekly basis. None of the respondents from small operations responded that they made a decision to fly when fatigued on a monthly basis.

4.3 Case and Control Comparisons

The primary significant differences found between Alaska commuter airline and air taxi cases and controls involved flight experience, duty hours, and bypass mail delivery. Case pilots reported fewer years of experience and half as many flying hours with their current employer as did control pilots. Pilots flying for case firms had half as many hours with their current employer as did control pilots. Alaska pilots for case operators flew, on average, 1 hour more per day and 10 hours more per week than control pilots, although fewer of them reported declining a flight due to fatigue. Case pilots were three times as likely as controls to fly daily into unknown weather conditions. Case operators were also significantly more likely to consider timely delivery of bypass mail important to their financial success.

Operators did not differ significantly in the risk reduction measures reported to be in place, although most measures were more commonly reported by case operators. However, the survey instrument could not distinguish whether risk reduction measures were historical or put in place in response to a crash. A high percentage of both case and control operators reported various training activities in place. No significant differences in training were found, despite the fact that several companies in the control group (but no cases) also held CFR Part 121 certificates and might be presumed to have more opportunity and support for in-house training than those without the additional certificate. Although reported training differed little between cases and controls, a higher percentage of case pilots reported deciding to fly into unknown weather on a daily basis compared to control pilots.

Differences between cases and controls could not be explained by case firms operating more aircraft, since data in the VIS indicated that case and control firms operated similar numbers of aircraft per operator. Differences between case and control companies regarding their fleet composition are unlikely to explain the differences observed in accident rates. According to NTSB data, single-engine aircraft, which predominated the fleet of case companies, do not have a higher fatal accident rate overall than rotorcraft, which were more common in control companies (1.22 accidents per 100,000 hours for single-engine aircraft, 1.48 accidents per 100,000 for rotorcraft[20]).

Although pilots in both groups worked long hours, a critical difference between pilots who flew for case operators and those who flew for control firms was time spent at work. The long hours are consistent with other surveys showing that pilots in regional flight operations worked longer duty days on average than in short-haul operations, even though their daily flight times were comparable.[21] Research has indicated that tired pilots flying short- and long-haul flights may not be aware of the effects of fatigue on their flight task performance.[22] The longer duty hours worked by case pilots combined with the higher rate at which they decide to fly into unknown weather may

Discussion

increase the risk of making a fatal error in judgment. Pilot age may also play a role, with pilot youthfulness acting as a factor mitigating the perception of fatigue.[22] This relationship should be investigated further to see if some pilots who operate under CFR Part 135—especially younger pilots—are flying such long hours, or so many hours per week, that they become unable to identify their limitations because of fatigue.

4.4 Strengths, Potential Limitations, and Biases

This survey is one of only a few studies in Alaska which has addressed safety practices and economic incentives that might put pressure on operators and pilots to fly in unsafe conditions. It is the only study in Alaska which randomly surveyed the entire air taxi and commuter airline industry, providing results which can be used to describe the state industry as a whole. The results will provide a baseline for future evolution of safety improvements associated with the Alaska Interagency Aviation Safety Initiative.

Survey bias can come from many sources, and all can limit the usefulness of the results. Biases can include a nonrepresentative sample, differences between those who respond to the survey and those who do not, and answers that are influenced by recent events and thus not representative of long-term attitudes and practices.

Every possible effort was made to select a representative sample that encompasses the diverse population of air taxi, commuter, and public agency operators and pilots across Alaska, and the results were weighted in accordance with the stratified sampling. Sample weights cannot correct for bias in who responds to a survey. Response bias means that people who responded to the survey may be different from people who did not participate. ISER research staff assessed response bias using publicly available data for the population. No significant relationship was found between accident rate and whether operators were willing to complete the survey. Likewise, size and location were not associated with a greater likelihood of response or refusal/noncontact. Other factors important to the analysis may have differed between respondents and nonrespondents; for example, operators with a limited concern for safety or who were experiencing financial difficulties may have systematically refused to respond. There is no way in this study to measure these possible effects; however, the 79% response rate appears high enough to suggest a representative sample.

Three events that occurred during the course of the operator survey may have affected many operators' responses. The tragic events of September 11, 2001, at the World Trade Center, Pentagon, and in Pennsylvania shut down aviation operations nationwide. In response to the uncertainty in the aviation industry and concern among respondents, interviewing was suspended for 1 week. Two serious air crashes in Alaska occurred during the following month. On October 10, 2001, one of the largest regional operators in Alaska sustained the worst commercial crash in Alaska since 1987—a crash on take-off that killed all 10 on board. On October 18, 2001, another of Alaska's largest regional carriers crashed a helicopter into Cook Inlet, killing three people. A series of events of this magnitude is likely to have affected operators' attitudes, perceptions, and business practices, but the extent of these effects is unknown.

5 | Recommendations

Based on the findings from this survey, Alaska operators should consider examining their work schedules and carefully evaluate whether their pilots have adequate opportunity for rest between duty times (or shifts) and sufficiently frequent days off to recuperate before returning to work. The National Aeronautics and Space Administration recommends sufficient breaks between duty periods to permit 8 hours of uninterrupted sleep in addition to sufficient time for transport, meals, and other essential activities.[23] Even the duty hours reported by control carriers in the case-control analysis may be too long to provide for adequate rest. During the busiest parts of the season, care should also be taken to minimize the impact of circadian shifts resulting from major changes in the timing of duty hours. Operators and pilots may also want to consider allowing time for naps and some physical exercise between busy duty periods.

Operators should also be informed of the potential risks associated with having inexperienced pilots flying long hours in Alaska. Although previous research is unclear as to the role inexperience plays, this analysis indicates that further research is needed to parse out how inexperience, especially inexperience flying in Alaska, contributes to fatal aviation accidents.

6 | Conclusion

Surveyed pilots' reported perception that their risk for fatal injury while working is low to moderate is not consistent with reality. The pilot fatality rate in Alaska is nearly five times the rate for all U.S. pilots (70 per 100,000 per year). Although survey results did not reveal a single underlying factor that could be identified as the cause of the high crash rates for Alaska commuter and air taxi operations, it did suggest that a number of factors—pilot fatigue and inexperience, financial pressures on operators, inadequate weather information—may be associated with higher accident risks. These factors may interact with each other to increase the risk. Policies and programs could address the factors and possibly mitigate the risks.

Fatigue has repeatedly been shown to be an important factor in increasing the probability of an aviation accident.[19, 23, 24] Current regulations (CFR Part 135.263, 135.265, 135.267, 135.269, 91.1057, and 91.1059) include rest requirements and flight hour limitations on a daily, quarterly, and annual basis. The duty hour limitations are only daily and do not directly impose weekly or monthly limits. Operators could adhere to these CFR's and still have pilots working up to 7 days of 14 hours a day or 98 hours a week. Total duty hours, especially for pilots flying long duty days, with many legs per day for many days in a row, are an important predictor of fatigue,[22, 25] and research by Goode showed that for CFR Part 121 operations, the relative proportion of accidents to exposure increases as the length of the duty day increases.[24]

The survey did not directly address why operators are having their pilots work such long duty days. Financial dependence on bypass mail may have created pressure for operators to keep pilots on long duty hours, possibly decreasing the margin of safety. Delivery of bypass mail involves flying in and out of small rural airports that have limited infrastructure and weather information services throughout the year. It is unlikely overtime pay was a motivator for pilots to work long hours, since few operators offered overtime pay. Instead, the short peak season and inevitable layoffs during winter may have provided an incentive for pilots to fly as much as possible when work was available.

Many of the responses received in these surveys were largely consistent with the objectives of three major programs in Alaska.

- Medallion Foundation: Pilot hazard training, which was supported by both operators and pilots in the survey, is being conducted by this foundation in a nonprofit, private-government collaboration to provide organizational tools, training, and technical support to the aviation industry in Alaska. The foundation, incorporated in November 2001, includes many of the procedural and training interventions mentioned in the survey and provides training opportunities to pilots in small operations.

- FAA's Circle of Safety: Announced to the public in late 2002, this program provides consumer education and emphasizes passenger understanding of the hazards of flying. While the industry expressed only guarded enthusiasm in the survey for passenger understanding as being a helpful intervention, clear consensus was expressed among survey respondents that passenger pressure to fly into adverse weather or poor visibility was an unhelpful and a potentially dangerous influence.

- Capstone: This program introduces high-technology navigational avionics in a compact suite of equipment designed for use in small aircraft. The first aircraft was equipped in November 1999. This innovative technology may provide much better information to inform pilot decision making and navigation. The latest Capstone proposal (phase III) is to equip all Alaskan aircraft with avionics (multifunction display for navigation, terrain, traffic and flight information; Automatic Dependent Surveillance-Broadcast and GPS Wide Area Augmentation System) and reduce dependency on earlier ground-based systems.[26]

These surveys have provided insight into the current operation of Alaska's aviation industry and assessed the acceptability of a wide range of possible interventions. Results from the survey and the additional case-control analysis might focus future research and prevention measures on identifying fatigue, stressing the importance of experience, and highlighting the potential costs of long days and work weeks for pilots.

References

1. National Transportation Safety Board. Aviation Accident Database. http://www.ntsb.gov/ntsb/query.asp. Accessed May 1, 2004.
2. National Institute for Occupational Safety and Health. *Worker Health Chartbook, 2004*. Cincinnati, OH: Department of Health and Human Services, Centers for Disease Control and Prevention; 2004. DHHS NIOSH 2004-146.
3. National Transportation Safety Board. *Aviation Safety in Alaska: Safety Study*. Washington, D.C.: National Transportation Safety Board; 1995. NTSB/SS-95/03.
4. Bensyl DM, Moran K, Conway GA. Factors associated with pilot fatality in work-related aircraft crashes, Alaska, 1990-1999. *American Journal of Epidemiology*. 2001;154(11):1037-1042.
5. Federal Aviation Administration. *Commuter Accidents in Alaska, 1990-1993*. Anchorage, AK: Federal Aviation Administration; 1993.
6. Federal Aviation Administration. *Proposed Action Plans for Reducing Alaskan Accidents in Fiscal Year 1993*. Anchorage, AK: Federal Aviation Administration; 1993.
7. Kobelnyk G. *A Review of Accidents Involving Alaskan Air Taxi Commercial Flight Operations 1983 to 2000*. Anchorage, AK: Federal Aviation Administration; 2004.
8. Middaugh JP. *The Epidemiology of Involuntary Injuries Associated with General Aviation in Alaska, 1963-1981*. Anchorage, AK: Alaska Division of Public Health; 1986.
9. National Transportation Safety Board. *Air Taxi Safety in Alaska*. Washington, D.C.: National Transportation Safety Board; 1980. NTSB/AAS-80/03.
10. Thomas TK, Bensyl DM, Manwaring JC, Conway GA. Controlled flight into terrain accidents among commuter and air taxi operators in Alaska. *Aviation, Space, and Environmental Medicine*. 2000;71(11):1098-1103.
11. American Airlines Training Corporation. *Definition of Alaskan aviation training requirements*. Anchorage, AK: Alaska Aviation Safety Foundation; 1982.
12. Bailey LL, Peterson LM, Williams KW, Thompson RC. Controlled flight into terrain: a study of pilot perspectives in Alaska. *Flight Safety Digest*. 2001;20(11/12):1-9.
13. Federal Aviation Administration. *Joint Interagency/Industry Study of Alaskan Passenger and Freight Pilots*. Anchorage, AK: Federal Aviation Administration; 1999.
14. Mondor C. Among U.S. states, Alaska has highest incidence of accidents in FARs Part 135 operations. *Flight Safety Digest*. 2001;20(11/12):43-50.
15. Federal Aviation Administration. *Audit of Alaska Air Carriers Focused on Take-off and Landing Accidents*. Anchorage, AK: Federal Aviation Administration; 1998.
16. Conway GA, Hill A, Martin S, et al. Alaska air carrier operator and pilot safety practices and attitudes: A statewide survey. *Aviation, Space, and Environmental Medicine*. 2004;75(11):984-991.
17. Thompson SK. *Sampling*. New York: John Wiley & Sons, Inc.; 1992.
18. Conway GA, Mode NA, Berman M, Martin S, Hill A. Flight safety in Alaska: comparing

attitudes and practices of high- and low-risk carriers. *Aviation, Space, and Environmental Medicine.* 2005;76(1):52-57.

19. National Transportation Safety Board. *Evaluation of U.S. Department of Transportation Efforts in the 1990s to Address Operator Fatigue.* Washington, D.C.: National Transportation Safety Board; 1999. NTSB/SR-99/01.
20. National Transportation Safety Board. *Annual Review of Aircraft Accident Data. U.S. General Aviation, Calendar year 1999.* Washington, D.C.: National Transportation Safety Board; 2003. NTSB/ARG-03/02.
21. Co EL, Gregory KB, Johnson JM, Rosekind MR. *Crew Factors in Flight Operations XI: A Survey of Fatigue Factors in Regional Airline Operations.* Moffett Field, California: National Aeronautics and Space Administration; 1999. NASA/TM-1999-208799.
22. Bourgeois-Bougrine S, Carbon P, Gounelle C, Mollard R, Coblentz A. Perceived fatigue for short- and long-haul flights: a survey of 739 airline pilots. *Aviation, Space, and Environmental Medicine.* 2003;74(10):1072-1077.
23. Rosekind MR, Neri DF, Dinges DF. From laboratory to flightdeck: Promoting operational alertness. *Fatigue and Duty Limitations? An International Review.* London: The Royal Aeronautical Society; 1997:7.1 - 7.14.
24. Goode JH. Are pilots at risk of accidents due to fatigue? *Journal of Safety Research.* 2003;34(3):309-313.
25. Rosekind MR, Boyd JN, Gregory KB, Glotzbach SF, Blank RC. Alertness management in 24/7 settings: lessons from aviation. *Occupational Medicine: State of the art reviews.* 2002;17(2):247-259.
26. Federal Aviation Administration. *Capstone Statewide (Phase III) Strategic Plan.* Anchorage, AK: Federal Aviation Administration; 2005.

Appendix A
Operator/Small Operator Questionnaire Summary

Appendix A. Operator/Small Operator Questionnaire Summary

The stratified sample meant that ISER staff spoke to almost all the larger operators in the state, but only about one in five of the smallest operators. To reflect the aviation community more accurately, the small operator results were weighted to account for the same two-thirds of all operators in the weighted sample that they represent in the total population. With the exception of OP1 and OP4, all counts are either to 153 or to the subset of 153 who responded to a given question. Occasionally, rounding the weighted results meant that the total answers varied by plus-or-minus 1. For OP1 (number of pilots) and OP4 (hours and departures), ISER staff estimated statewide totals representing the estimated 345 operators active in Alaska during the survey period.

OP1. How many pilots do you currently employ?
Average of 5.38 pilots with a range of 0 to 105 pilots currently employed.
Just over half (52%) of Alaska operators employ only 1 pilot.
Pilot estimate for all Alaska companies: 1856 +/- 185

OP2. How many pilots do you typically employ each season?
Averages:

| **5.57** | Summer | **4.75** | Autumn | **4.12** Winter | **4.52** Spring |

OP3. How many pilots did your company hire in each of the last 3 years?
Average:

1.78	2001 (total expected)
1.94	2000
1.76	1999

OP4. Please list total flight hours and departures flown by your company in 2000 (both scheduled and unscheduled). Include all locations and aircraft.
This estimate is for all operators statewide.

| 420,000 (275,000-565,000) | Scheduled flight hours | 412,000 (293,000-532,000) | Scheduled departures |
| 415,000 (370,000-460,000) | Unscheduled flight hours | 503,000 (405,000-594,000) | Unscheduled departures |

OP5. Every year the FAA mails a survey–the General Aviation and Air Taxi Activity survey– to some aircraft owners asking about how the aircraft was used during the previous year. Has your company received any of these surveys the last three years? (You may have received several surveys asking about different aircraft.)
27%: No   48%: Yes   25%: Don't Know
↓
How many were received, and how many were completed and returned?

Year	Number of surveys received	Number of surveys completed / returned
2001	**Average: .97**	**Average: .56**

| 2000 | Average: 1.23 | Average: .99 |
| 1999 | Average: .97 | Average: .80 |

OP6. If you had to choose between the types of experience listed below when hiring a pilot, which would be the most important? Which would rank second? 3rd? 4th? [These questions were asked of only of the 64 (weighted) companies that had hired pilots.]

[# = Respondents]

	1st	2nd	3rd	4th	No Answer
a. Flying in the area of Alaska where the pilot will be employed	30	13	7	11	2
b. Flying anywhere in Alaska (total Alaska flight hours)	2	14	29	17	2
c. Total Flying hours anywhere	14	11	11	26	3
d. Flying in the type of aircraft your company uses:	16	24	14	8	2

OP7. Is there some other pilot experience or qualification more important than any of these four?

No (48% of operators)   Yes (52% of operators)
⬇
OP7a. Please describe it.

o Specific character qualities were most commonly cited as important. Qualities listed were: Judgment, decision-making, attitude, discipline, steady, level head, professional, cautious, commitment to safety, older, mature.
o Experience was the second most often listed qualification. Types of experience listed include: Alaska specific, large aircraft, interagency, firefighting, flight instructor, owner flown, off-airport, external load, seaplanes, round engine, bad weather, remote area, and flying IMC w/out radar
o A few listed the class of aircraft as important.
o A few listed accident/incident face hours/violation face hours as important.

OP8. How much of a problem is pilot fatigue in pilot scheduling?

Major problem: 7
Minor problem: 58
Not a problem: 87
(1 missing answer)

Appendix A. Operator/Small Operator Questionnaire Summary

OP9. How do you pay your pilots? (Not asked of 1-pilot operators)

Hourly for all duty hours:	2
Hourly for flight hours only:	10
Salary:	21
Combination of salary and flight hours:	25
Combination of flight hours, duty hours and salary:	7
Flight completions:	None
Other (please explain):	9

- **Most explained some minimum salary plus extra pay for extra hours, days, etc.**
- **Some pay by hourly rate with different rates for flight hours, standby hours, and extra hours.**
- **Some said that they pay by daily rate.**
- **A few pay per customer as part of the wage**
- **One listed family labor**

OP10. Do you pay your pilots overtime? (Not asked of 1-pilot operators)

58 No   **16** Yes   (out of 73 Respondents)

OP10a. Under what conditions?

- **Half of those who pay overtime do so for all hours worked over some fixed limit. The standard ranged from 60 hours per month to 100 hours per month; from 4 flight hours to 8 total hours per day.**
- **Several paid extra for pilots temporarily stationed away from home.**
- **A few paid overtime for extra days worked, such as on scheduled days off, or for "hard work."**

OP11. From your personal experience, are the Federal Aviation Regulations interpreted consistently by different inspectors at different times?

67 No   **86** Yes   1   No Answer

OP11a. Please give one or more examples.

- **Operators deal with multiple inspectors and say that each inspector interprets the regulations differently.**
- **They listed maintenance requirements, paperwork, airworthiness, service directives, and operations procedures, flight and duty time rules and de-icing rules.**

OP12. Do you feel that some Federal Aviation Regulations interfere with getting the job done, without contributing to safety?

63 No   **86** Yes   **4** No Answer
⬇

OP12a. Please indicate which regulations and why (in order of frequency):

- Visibility, weather and altitude regulations
- Hazardous materials regulations
- Duty time regulations
- Interpretation of regulations (rather than the regulation itself)
- Drug testing.
- Maintenance regulations.

OP13. Does your company require higher than FAA weather minimums for flying?

118 No   **33** Yes   **3** No Answer
⬇

OP13a. Please of describe your policy or attach a copy. Note when your company began this requirement

- No one attached a copy of their policy; most wrote a brief description
- New or inexperienced pilots are held to higher weather minimums
- Greater visibility required at night or in the mountains
- Several said they simply don't fly in bad weather
- When implemented: Answers range from 1970 to 2001

OP14. Does your company have written programs for **pilot training** to help pilots deal with the following conditions: [# = **Respondents**]

	Yes	No	If yes, when started
a. Whiteout conditions	41	109	1970-2000
b. Low visibility conditions	49	102	1970-2000
c. Flat lighting conditions	41	110	1970-2000
d. Recovery from inadvertent flight into IMC	51	99	1970-2001

OP15. Does your company have written programs for **pilot checking** to ensure pilot proficiency in the following conditions:

	Yes	No	If yes, when started
a. Whiteout conditions	32	119	1975-2000
b. Low visibility conditions	38	113	1970-2000
c. Flat lighting conditions	31	120	1975-2000
d. Recovery from inadvertent flight into IMC	41	110	1970-2001

Appendix A. Operator/Small Operator Questionnaire Summary

OP16. Who can decide to cancel a flight?

Pilot (Yes: 151, No: 1, Missing: 1)
Someone else in the company (Yes: 64, No: 72, Missing: 17)
↓
(Who is it?) **Common answers: Chief Pilot, Director of Operations, Dispatch, Manager, Owner, and President**

Someone outside the company (Yes: 14, No: 122, Missing: 17)
↓
(Who is it?) **Common answers: Customer/Passenger**

OP17. If a non-pilot employee makes decisions about launching flights, what training (initial and recurrent) does the company provide or require that person to have?

- **Some (especially those referring to passengers canceling) said that no training is provided or required.**
- **Others cited some type of dispatch training, including weather training.**
- **Several listed informal training, such as discussions, experience, working with the operations manager.**
- **A few stated they train personnel using the operations manual procedures or the Federal Aviation Regulations.**
- **A few cited training without specifying the topic(s).**

OP18. Does your company have a written list of required conditions to launch a flight (for example, a risk assessment worksheet)?

143 No   **10** Yes
                  ↓
              OP18a   When did your company start using this list?
                       Ranged from 1971-2001

              OP18b.   Please attach a copy of this list.
                       Only 3 included a list

OP19. How many of your aircraft have the following types of equipment?

Number of operators with each type of equipment in none, or some, of their aircraft:	0 aircraft	1 or more aircraft	No Answer
a. Auto pilot	97	51	5
b. Very high frequency Omni-directional Range	39	111	3
c. Global Positioning System–Visual Flight Rules	9	141	4
d. Global Positioning System–Instrumental Flight Rules	98	49	6
e. Long range navigation	116	31	6
f. Mid-air collision avoidance system	135	13	6
g. Other Avionics:	4	66	83
h. Pilot shoulder harness	7	142	4
i. Rear Passenger shoulder harness	91	58	4
j. Pilot 5-point restraint harness	122	26	6
k. Other crash protection equipment	131	19	3

Please rate each type as very helpful, somewhat helpful, or not at all helpful **to flight safety in Alaska** (not just to your company). [**# = Respondents**]:

Type of equipment	Very Helpful	Somewhat Helpful	Not at all Helpful	Missing
a. Auto pilot	30	28	23	72
b. Very High Frequency Omni-directional Range	51	50	18	34
c. Global Positioning System–Visual Flight Rules	132	7	2	12
d. Global Positioning System–Instrumental Flight Rules	48	19	10	76
e. Long Range Navigation	9	22	40	83
f. Mid-air collision avoidance system	24	16	17	96
g. Other Avionics:	50	23	1	80
h. Pilot shoulder harness	122	14	6	12
i. Rear Passenger shoulder harness	49	29	5	70
j. Pilot 5-point restraint harness	21	29	10	93
k. Other crash protection equipment:	18	4	1	130

Appendix A. Operator/Small Operator Questionnaire Summary

OP20. How important is each of the following to your **company's financial success**?
[**# = Respondents**]

	Very Important	Important	Not Important	Not Applicable	No Answer
a. On-time delivery of mail	15	13	23	97	5
b. On-time delivery of passengers	50	61	18	23	1
c. On-time delivery of cargo	36	57	26	29	6

OP21. In the last 18 months, have your company's insurance costs per seat changed?

    112 Yes   **38** No

    106 Increased
      6 Decreased
    ↓

OP21a. By what percent did they increase or decrease?
Increases ranged from 1% to 320%
Mean: 23%
Median: 15%
Mode: 10%

OP21b. If your insurance costs changed, why do you believe they changed?
- **Accident rates, court claims, industry losses were most frequently cited.**
- **Next most frequent was lack of competition, too few insurance companies, greed.**
- **Other reasons noted once were inflation, September 11th.**

OP22. What survival equipment, beyond legally-required items, is in your company aircraft?

- **About one-third said that they carry only what is legally required.**
- **Others listed sleeping/camping gear such as blankets, sleeping bags, tarps, and tents, food and/or water, stove, extra clothing, ax, knife, signaling devices, a company survival kit, and/or a first aid kit**
- **Some listed satellite phones, and/or cell phones**
- **One carries a life vest; one a life raft.**

OP23. What training does your company provide to use the survival equipment in the aircraft?
- **One in four operators said they provide no training, just self-training, or pilots "should know."**
- **Most of those who provide training incorporate it into their regular pilot training.**

- Some listed a type of informal training such as a discussion, staff meeting, or training by a manual.
- A few offer annual, intensive, crash-survival training such as egress, water egress, and fire control training.

OP24. The table below asks your opinion about measures that might improve aviation safety throughout Alaska (not just in your company). For each measure, rate **how effective** you think it could be in preventing aircraft crashes if it were widely applied in Alaska aviation. [# = **Respondents**]

Possible measures to use in preventing aircraft crashes	Very Effective	Somewhat Effective	Not Effective
Pilot training improvements in the following areas:			
a. Meteorology	68	67	14
b. Decision-making	105	36	8
c. White-out/flat-light conditions	85	58	5
d. Regional hazards	103	40	5
Company policies and procedures			
e. Written criteria for go/no-go decisions	46	61	39
f. Rewards from management for flights or flight hours without accidents/incidents	34	66	46
g. Pay based on salary rather than flight hours or flights	43	73	30
h. More flight time required of new pilots	59	54	33
i. Better checks of a pilot's flying history before hiring	45	62	39
j. More locations with manned weather reporting	127	23	2
k. More locations with automated weather reporting	90	48	13
l. Increased accuracy of existing weather reporting	117	31	4
m. Increased and improved use of video cameras, such as mountain pass cameras	107	30	12
n. Improved passenger understanding of weather hazards	48	54	48
o. Changes in how by-pass mail is given to operators	16	28	71
p. More time to deliver by-pass mail before it's switched to another operator	18	32	64
q. Financial incentives (e.g., lower insurance rates, preference in mail contracts) for flights or flight hours without accidents/incidents	58	46	38

Appendix A. Operator/Small Operator Questionnaire Summary

OP25. If you had to choose just two of the above items as most useful, which would they be? Please indicate the appropriate letter.

Top 5 Answers	# citing as important
24J: More Locations with weather reporting	76
24B: Pilot training in decision making	45
24D: Pilot training in regional hazards	27
24M: Increased use of video cameras (mountain pass cameras)	27
24L: Increased accuracy of existing weather reporting	25

OP26. Are there any other measures we didn't mention that might improve aviation safety in Alaska?

37 Yes   **14** No   102 Missing

What are they?

- **More experience and more training for pilots most commonly cited.**
- **Many listed technical improvements, such as Global Positioning System, Capstone avionics, more weather cameras.**
- **Several wanted more manned weather stations.**
- **Some cited a good pilot attitude, common sense, and a commitment to safety.**
- **A few said that weather and visibility regulations need to be enforced, as well as the Visual Flight Rules minimum increased.**
- **Other reasons noted included a shorter duty day, more flexible Federal Aviation Regulations, more knowledgeable management.**

OP 27. Please add any other comments about aviation safety in Alaska you think we should know.

- **Change the attitude of pilots flying in Alaska.**
- **Education of the customer, if customer pressure is mitigated that is one safety initiative.**
- **Major concern with maintenance, maintenance guys aren't getting paid enough and are overworked.**
- **One of the biggest driving forces for safety is insurance companies, Federal Aviation Administration doesn't have any teeth at all.**
- **Don't do these surveys during peak season.**
- **Real shortage of competent pilots available for seasonal work in AK. Forced to hire pilots below standards. No amount of training in a one or two-week period is going to help these pilots.**
- **Require higher than average pilot experience, 3 times Federal Aviation Administration minimum 1500 hours, can't hire teenagers as some carriers do.**
- **Federal Aviation Administration inspectors are good, but their caseloads are too**

large; they don't have time to help companies make improvements.
- Training pilots not to push weather with aircraft that are Visual Flight Rules-equipped, that it is all right to turn around or stop the flight.
- Need pilot license in State in which you're flying including proficiency in actual flight area.
- One Question not asked is: Should the Federal Aviation Administration up the drug and alcohol policy? What is the ratio of private pilot to company pilot fatalities?
- More cameras/Capstone!
- No time builders, should want to fly and stay in Alaska.
- Safety needs to start at the top within each company and be effectively communicated throughout the company.
- New pilots from outside need close supervision and decision-making help until they gain experience.
- All aircraft should be equipped for flight icing conditions if used in air carrier service. All air carrier flights should be Instrument Flight Rules all the time.
- Needs some accountability for Federal Aviation Administration inspectors when they interpret regulations.
- Get manned weather into all villages. That will cut down on accidents 25%.
- Pilot judgment errors are the principal cause of accidents.
- Automated weather systems are not working.
- Flight Standards District Office often contributes to pilots flying in poor weather by allowing Visual Flight Rules only operators to fly when weather is below Visual Flight Rules min (500 +2).
- Need to have a committee (union, etc.) to represent pilots who need specific help in solving aviation-related needs.

Appendix A. Operator/Small Operator Questionnaire Summary

Appendix B
Pilot Questionnaire Summary

Appendix B. Pilot Questionnaire Summary

Throughout, MV = N indicates the number of missing values for the question.

I would like to begin by asking you a few questions about your flying career and your background. **(MV = 3)**

A1. Which pilot ratings and certificates do you hold?

Commercial **Yes = 210; No = 50**   Single-engine land **Yes = 218; No = 42**
Instrument **Yes = 195; No = 65**   Multi-engine land **Yes = 156; No = 104**
ATP **Yes = 132; No = 128**   Single-engine sea **Yes = 142; No = 118**
Helicopter **Yes = 50; No = 210**   Multi-engine sea **Yes = 39; No = 221**
Flight instructor **Yes = 73; No = 187**   Others (please specify) **Yes = 25; No = 235**
↓

A&P Mechanic, Airframe/powerplant, Airframe mechanic, BV234, CFI Instrument, Commercial Sea, CU580, C212, DC3, DC6, DC7
Flight engineer, Glider, Glider/sail plane, Gliding AeroTow, LR-Jet/glider, Mechanical airframe, Multiple others, Tail Wheel time, Turbojet engineer, Type SK70

(Of the pilots who answered "no" to commercial, 40 had an airline transport pilot license, 8 were public agency employees [who don't need a commercial] license, and 2 were Part 135 pilots who said they didn't have a commercial license.)

A2. How many total hours have you flown in Alaska? How many in the last 12 months? Now can you tell us how many Alaska departures have you made in your total flight career? In the last 12 months? Finally, can you tell us how many total hours you have flown in all locations, including Alaska? How many in the last 12 months?

Flight Hours	**Alaska**	**Alaska Departures**	**All Locations, including Alaska**
Total Flight Career	N=261; MV=2; Mean=7374.78	N=187; MV=76 Mean=12480.96	N=256; MV=7; Mean=9340.6
Last 12 months	N=261; MV=2; Mean=633.66	N=186; MV=77; Mean=1445.18	N=245; MV=18; Mean=664.84

The next questions ask about your total flight career.

A3. How many instrument hours have you flown in Alaska? ➔ **N=260; MV=3; Mean=567.83 Median=77.5**

　　A3a. Can you estimate your total number of instrument hours?
　　N=259; MV=4; Mean=885.505; median =175

A4. How many hours have you flown for your current employer?

N=202; MV=61; Mean=3542.104

A5. Thinking about the number of years you have flown in Alaska, how many have been seasonal?
N=199; MV=64; Mean=2.256 Years

    A5a.   How many have been year-round? **N=202; MV=61; Mean=9.244 Years**

A6. Please estimate what percent of your paid flight hours in 2000 occurred in each season. (For example, 100% summer; or 25% spring, 50% summer, 25% autumn, etc.)

%	Spring		%	Autumn
%	Summer		%	Winter

A7. Over your entire career, how many different companies have you worked for as a pilot?
N=262; MV=1; Mean=3.90

    A7a.   Over how many years has that been?
N=258; MV=5; Mean=16.076

A8. What is your gender? **(MV=1)**
Female: **N= 13**
Male: **N=249**

    A8a. How old are you? **(MV=3)**
N=260; Mean=44.08

A9. What is your race? **(2 refused; MV=2)**

American Indian or Alaska Native: **10**	Asian Indian: **0**
White: **251**	Japanese: **0**
Black, African American, or Negro: **0**	Native Hawaiian: **0**
Chinese: **0**	Guamanian or Chamorro: **0**
Korean: **0**	Filipino: **0**
Vietnamese: **0**	Samoan: **0**
Other Asian: **1**	Other Pacific Islander: **0**
Some other race (please specify): **1**	

↓
Other described: Latino=1

A10. What is the highest level of formal education you have completed? (PLEASE MARK ONLY ONE) **(MV=2)**

Attended high school; didn't graduate: **2**   Associate's degree: **33**

Appendix B. Pilot Questionnaire Summary

        GED: **3**   Bachelor's degree: **95**
        High school diploma: **33**   Master's degree: **13**
        Attended college; no degree: **81**   Doctoral degree: **1**

A11. Now, based on your experience as a pilot, do you feel that some Federal Aviation Regulations interfere with getting the job done, without contributing to safety? **(MV=10)**

        No **149**   Yes **104**
                     ↓

        A11a. Can you give me one or more examples? **(examples given=94)**

A12. Are there routes, locations, or conditions that should require higher than FAA minimum weather conditions for flying? **(MV=67)**

        No **148**   Yes **48**
                  ↓

        A12a. What are they? **(Descriptions included=48)**

A13. During the peak season, what hours do you typically work each day, including periods of time you are not on duty?

From _____ AM/PM to _____ AM/PM
On average from 0700 to 1900; calculated mean hours/day=11.8; N=261; MV=2

A14. During the peak season, how many **hours per day** are you typically **on duty**?
Duty hours per day 11.931; N=260; MV=3

A15. During the peak season, how many **days per week** do you typically work?
Days per week mean=5.897; N=261; MV=2

    7 days   **89 pilots**
    6 – 6.5   **107**
    5 - 5.5   **45**
    <5   **11**
    No typical week   **9**

A16. During the peak season, how often would you have liked to decline a flight due to fatigue, but you flew anyway? (MARK ONLY ONE) **(MV=2)**

    Daily: **2**   Less often than monthly: **51**
    Weekly: **20**   Never: **172**
    Monthly: **16**

A17. How often do you have to decide whether to fly into unknown weather conditions that may deteriorate below Visual Flight Rules minimums? (MARK ONLY ONE)
(MV=4; Don't know=1)
- Daily: **23**
- Weekly: **64**
- Monthly: **63**
- Less often than monthly: **67**
- Never: **41**

A18. How often do you fly into weather that is different from what was predicted when you started your flight? (MARK ONLY ONE) **(MV=1)**

- Daily: **22**
- Weekly: **108**
- Monthly: **7**
- Less often than monthly: **47**
- Never: **6**

A19. How often do Flight Service Stations provide accurate, current weather conditions for where you fly? (MARK ONLY ONE) **(Don't use = 2; Don't know = 1; MV = 2)**

- Always: **15**
- Most of the time: **202**
- Occasionally: **22**
- Rarely: **14**
- Never: **5**

A20. While working for your current employer, have you declined a flight due to poor visibility or other weather-related reasons? **(MV=61)**

No: **16**  Yes: **186**

A21. Did the company support your decision? **(Also: Not always=1; Missing Answer=81)**

No: **179**   Yes: **2**

A22. Do you have standard procedures to follow if you unexpectedly fly into IMC?

No: **63**   Yes: **196**   **(MV = 4)**

A23. Has your employer provided you with training and/or check rides to help you deal with **white-out conditions**? **(MV=2)**

	Training	Check Rides	Neither	Both
a. White-out conditions (MV=65)	83	6	39	70

Appendix B. Pilot Questionnaire Summary

b. Low visibility conditions? (**don't know=1; MV=64**)	82	7	21	88
c. Flat light conditions? (**MV=64**)	83	7	39	70
d. Recovery from inadvertent flight into IMC? (**MV=66**)	77	8	27	85

A24. How confident are you that you can safely fly under Visual Flight Rules in **low visibility conditions**? Are you very confident, somewhat confident, or not confident?

	Very confident	Somewhat confident	Not confident	Missing Answer
a. Low visibility	210	40	5	8
b. Flat-light conditions	185	51	16	11
c. White-out conditions (Don't know = 1)	164	44	43	11

A25. What survival training have you received from your current employer?
Training received = 179; None/No training received=21; MV=63

A26. From the list of resources I am going to read, which ones do you use when making the decision to launch a flight? (CHECK ALL THAT APPLY)

Flight Service Station
Yes=236; No=26; MV=1

National Weather Service
Yes=220; No=42; MV=1

Station Manager or other company personnel at destination(s)
Yes=206; No=56; MV=1

Dispatcher, flight follower, other company personnel at hub or headquarters **Yes=220; No=42; MV=1**

AWOS/ ASOS **Yes=232; No=30, MV=1**

Pilots who are in route or who have flown the route that day **Yes=229; No=4; MV=30**

Other (please specify) *(see list below)*
Yes = 69; No = 193; MV = 1

Agents in villages, All available, Auto Advisory Frequency, Aviation weather, National Oceanic and Atmospheric Administration, and more, Boats/ships, Call destinations, Capstone weather feature, Chief Pilot, Company wind charts, Community people you know, Company policies, Current weather/forecasts, Customers in field, Destination weather, Direct User Access Terminal, Federal Aviation Administration website, General observations, International weather package provided, Internet, Internet cameras, Internet weather information, Internet/Marine weather, Juneau 800 weather number, Local people, Look out my window, Marine weather station, National Oceanic and Atmospheric

Administration, Observers in villages, Others along the way, Others at destination, People around state, People at destination, People in field, Personal observation, Phone people in field, Pilot judgment of conditions, Pilot observation, Radar, Remote weather observations, Sparrvon Village, Station personnel, Unofficial weather observer, Video cameras, Visual observation, Weather cameras, Web cameras, Weather printout.

A27. If you refuse to launch a flight due to marginal weather, how likely is it that your passengers will fly with a different company? Is it not at all likely, somewhat likely, or very likely?

	Not at all likely	Somewhat likely	Very likely	Don't know	Not applicable
a. Your passengers will fly with a different company? **MV = 2**	142	65	20	7	27
b. The Post Office will give bypass mail to another company? **MV = 3**	73	15	9	23	140
c. Some other pilot will comment that they could have completed the flight? **Refused= 1; MV =3**	155	35	10	20	39

A28. Compared to other jobs, how safe is your pilot job? Is it much safer than other jobs, slightly safer, as safe as other jobs, slightly more dangerous, or much more dangerous than other jobs? **(Don't know= 1; MV = 4)**

Much safer than other jobs: **17**
Slightly safer than other jobs: **26**
As safe as other jobs: **108**
Slightly more dangerous than other jobs: **91**
Much more dangerous than other jobs: **16**

A29. Do you have any accidents or incidents on your record? **(MV=3)**

No: **205**   Yes: **55**

A30. Now I'm going to read some different types of avionics, and I would like you to tell me how helpful you think each is in preventing crashes? How helpful is the **auto pilot**? Is it very helpful, somewhat helpful, or not helpful? (MARK ONE ANSWER FOR EACH)

Appendix B. Pilot Questionnaire Summary

	Very Helpful	Somewhat Helpful	Not Helpful	Don't know	Missing Answer
a. Auto pilot	68	88	50	22	35
b. Very High Frequency Omni-directional Range	70	128	41	3	21
c. Global Positioning System – Visual Flight Rules	211	31	11	3	7
d. Global Positioning System – Instrument Flight Rules	128	61	19	18	37
e. Long Range Navigation	6	49	136	35	37
f. Mid-air collision avoidance system	117	45	25	33	43
g. Other avionics (**No other avionics mentioned = 56**) -- see list below --	148	23	2		34

Automatic Direction Finder (ADF); Automatic Direction Finder transponder; Automatic Direction Finder, radar altimeter, radar; Automatic Direction Finder, Transponder; Automatic Direction Finder/Non-directional Beacon (NDB); Automatic Direction Finder/Very High Frequency; Automatic Direction Finder-Direction Finder-radar; Automatic Direction Finder-Satellite phone; Automatic Direction Finder-weather-radar-mapping feature; Air cell phone; Capstone; Data link/moving map; District Measuring Equipment (DME); District Measuring Equipment localizer; District Measuring Equipment-Automatic Direction Finder; Double all avionics; Dual Global Positioning System/Communications; Dual Very High Frequency; Early Warning Systems; Electric Flight Instrument System (EFIS); Electric Flight Instrument System-Radar Terrain System; Electric Flight Instrument System; Flight Management System, Terrain Awareness and Warning System (TAWS); Global Positioning System (GPS); Ground Proximity Box; Hand-held Global Positioning System; Handheld Global Positioning System; High-frequency radio; Horizontal Situation Indicators (HSI), Global Positioning System moving map; Illuminator-Radio altimeter, Maximum Working voltage; Marine meter; Marine band Very High Frequency (VHS); Marine VHF radio; Missing answer; Moving 3D map; Non-Directional Beacon (NDB); NDB-Automatic Direction Finder; Personal Locator Beacon; Programmable VHF; Radar-weather; Radar; Radar altimeter; Graphical Weather Systems; Radar illuminator; Radio; Radios; Radio Magnetic Indicator-Flight Director, Moving map; Satellite phone; Satellite phone, flight director; Satellite phone, GPS; Satellite phone, hand-held GPS; Standard communications equipment; Traffic alert and Collision Avoidance System (TCAS); Terrain map, Capstone MX20; Traffic alert and Collision Aviation System, ground surveillance; Transponder; Transponder, Satellite phone; Transponder-Automatic Direction Finder; Very High Frequency Aviation Transmission/Marine Transmission; Very High Frequency radio; Wing leveler; Weather radar/radio altimeter.

Appendix B. Pilot Questionnaire Summary

A31. I would like to ask your opinion about measures that might improve aviation safety for all pilots in Alaska.

Can you tell me how effective **pilot training improvements in meteorology** could be in preventing aircraft crashes if widely applied in Alaska aviation? Would it be very effective, somewhat effective, or not effective?

Possible measures to use in preventing aircraft crashes	Very Effective	Somewhat Effective	Not Effective	Don't Know
Pilot training improvements in the following areas:				
a. Pilot training improvements in meteorology (MV = 5)	144	93	19	2
b. Pilot training improvements in decision-making (MV = 5)	192	55	10	1
c. Pilot training improvements in white-out/flat-light conditions (MV = 6)	185	59	12	1
d. Pilot training improvements in regional hazards (MV = 5)	206	44	7	1

Now I'm going to ask about company policy and procedures, would company policies that included **written criteria for Go/No Go decisions** be very effective, somewhat effective, or not effective in preventing aircraft crashes?

	Very Effective	Somewhat Effective	Not Effective	Don't Know
Company policies and procedures				
e. Written criteria for go/no-go decisions (MV = 7)	89	121	45	1
f. Rewards from management for flights or flight hours without accidents/incidents (MV = 7)	59	137	59	1
g. Pay based on salary rather than flight hours or flights (MV = 8)	100	125	28	2
h. More flight time required of new pilots (MV = 7)	107	105	42	2
i. Better checks of a pilot's flying history before hiring (MV = 7)	53	135	67	1

Now thinking about the weather, would more locations with **manned** weather reporting be very effective, somewhat effective, or not effective in preventing aircraft crashes?

Air Transportation Safety in Alaska

Appendix B. Pilot Questionnaire Summary

	Very Effective	Somewhat Effective	Not Effective	Don't Know
Weather				
j. More locations with manned weather reporting (**MV = 6**)	210	41	6	
k. More locations with automated weather reporting (**MV = 6**)	186	59	12	
l. Increased accuracy of existing weather reporting (**MV = 6**)	201	50	6	
m. Increased and improved use of video cameras, such as mountain pass cameras (**MV = 7**)	197	40	16	3
n. Improved passenger understanding of weather hazards (**MV = 6**)	47	120	90	

I'll move on now to operating environments for companies like yours. Would changes in **how by-pass mail is given to operators** be very effective, somewhat effective, or not effective in preventing aircraft crashes?

	Very Effective	Somewhat Effective	Not Effective	Don't Know
Operating Environment				
o. Changes in how by-pass mail is given to operators (**MV = 41**)	19	74	119	10
p. More time to deliver by-pass mail before it's switched to another operator (**MV = 44**)	22	82	106	9
q. Financial incentives (e.g., lower insurance rates, preference in mail contracts) for flights or flight hours without accidents/incidents (**MV = 18**)	54	109	80	2

A32. Thinking of those 17 measures you just rated, if you had to choose only two as most useful, which would they be?

Most frequent selections:

1. a31b=80; a31d=41; a31j=39; a31a=35
2. a31j=75; a31h=21; a31k=21; a31m=20; a31l=19

A33. **If there are other measures that you believe might improve aviation safety in Alaska, but which we didn't discuss in the previous question, can you tell me what they are? Comments included=176; None/NA=84, MV=3**

34. Do you have any other comments you would like to add about aviation safety in Alaska? **Comments included=177; MV=86**

Capstone Pilot Module Screening

Have you ever flown Capstone equipped aircraft for your company?

○   No →   Thank you. You do not need to complete the Capstone module.

○   Yes →   [Please turn page and continue.]

Thank you for your time. All of the information you have provided is confidential and cannot be used for enforcement purposes.

Appendix B. Pilot Questionnaire Summary

Appendix C
Additional Resources

Contact information is provided for three of the programs which promote safety in Alaska's commercial aviation Community.

Capstone Program
http://www.alaska.faa.gov/capstone/Index.htm
Capstone Program Office
Sue Gardner, Program Manager
801 B Street, Suite 300
Anchorage, Alaska 99501
Phone: (907) 271-1338

Federal Aviation Administration's Circle of Safety
http://www.alaska.faa.gov
U.S. Department of Transportation
Federal Aviation Administration
Office of Community Relations
222 West Seventh Avenue
Anchorage, Alaska 99513-7587
Phone: (907) 270-5296

Medallion Foundation
http://medallionfoundation.org
2301 Merrill Field Drive, Suite A3
Anchorage, AK 99501
Phone: (907) 222-3210
Fax: (907) 222-3206

www.ingramcontent.com/pod-product-compliance
Lightning Source LLC
Chambersburg PA
CBHW081739170526
45167CB00009B/3879